AF880986

Recreational UAV - Drone

Pocket Logbook

Tried and Trusted Indie Publishing

ISBN 978-1-925332-38-4 (Soft Cover Edition)

Disclaimer

The author and/or publisher is not rendering legal or accounting opinions and cannot accept responsibility for statements contained within this book.

This publication is intended for private record keeping. Those wishing to use it for taxation or business records should ensure they know the current requirements in their operating jurisdiction.

Tried and Trusted Indie Publishing
www.tatindiepublishing
Melbourne, Australia

Drone Details

Drone Make:

Drone Model:

Description:

Modifications:

Owner

Name:

Address:

City: State:

Postcode:

Phone No:

FLIGHT RECORD

YEAR		AIRCRAFT		FLIGHT TIME					GPS OR MAP LOCATION/ROUTE/REMARKS
									Totals Brought Forward
Month	Day	Type	Model	Start	End	Hours	Take Offs	Landings	
Signature									Totals to carry over

FLIGHT RECORD

YEAR		AIRCRAFT		FLIGHT TIME					GPS OR MAP LOCATION/ROUTE/REMARKS
									Totals Brought Forward
Month	Day	Type	Model	Start	End	Hours	Take Offs	Landings	
Signature									Totals to carry over

FLIGHT RECORD

YEAR		AIRCRAFT		FLIGHT TIME					GPS OR MAP LOCATION/ROUTE/REMARKS
									Totals Brought Forward
Month	Day	Type	Model	Start	End	Hours	Take Offs	Landings	
Signature									Totals to carry over

YEAR		AIRCRAFT		FLIGHT TIME					GPS OR MAP LOCATION/ROUTE/REMARKS
									Totals Brought Forward
Month	Day	Type	Model	Start	End	Hours	Take Offs	Landings	
Signature									Totals to carry over

FLIGHT RECORD

YEAR		AIRCRAFT		FLIGHT TIME					GPS OR MAP LOCATION/ROUTE/REMARKS
									Totals Brought Forward
Month	Day	Type	Model	Start	End	Hours	Take Offs	Landings	
Signature									Totals to carry over

FLIGHT RECORD

YEAR		AIRCRAFT		FLIGHT TIME					GPS OR MAP LOCATION/ROUTE/REMARKS
									Totals Brought Forward
Month	Day	Type	Model	Start	End	Hours	Take Offs	Landings	
Signature									Totals to carry over

FLIGHT RECORD

YEAR		AIRCRAFT		FLIGHT TIME					GPS OR MAP LOCATION/ROUTE/REMARKS
									Totals Brought Forward
Month	Day	Type	Model	Start	End	Hours	Take Offs	Landings	
									Totals to carry over

Signature ______________________

FLIGHT RECORD

YEAR		AIRCRAFT		FLIGHT TIME					GPS OR MAP LOCATION/ROUTE/REMARKS
									Totals Brought Forward
Month	Day	Type	Model	Start	End	Hours	Take Offs	Landings	
Signature									Totals to carry over

FLIGHT RECORD

YEAR		AIRCRAFT		FLIGHT TIME					GPS OR MAP LOCATION/ROUTE/REMARKS
									Totals Brought Forward
Month	Day	Type	Model	Start	End	Hours	Take Offs	Landings	
Signature									Totals to carry over

FLIGHT RECORD

YEAR		AIRCRAFT		FLIGHT TIME					GPS OR MAP LOCATION/ROUTE/REMARKS
									Totals Brought Forward
Month	Day	Type	Model	Start	End	Hours	Take Offs	Landings	
Signature									Totals to carry over

FLIGHT RECORD

YEAR		AIRCRAFT		FLIGHT TIME					GPS OR MAP LOCATION/ROUTE/REMARKS
									Totals Brought Forward
Month	Day	Type	Model	Start	End	Hours	Take Offs	Landings	
Signature									Totals to carry over

FLIGHT RECORD

YEAR		AIRCRAFT		FLIGHT TIME					GPS OR MAP LOCATION/ROUTE/REMARKS
									Totals Brought Forward
Month	Day	Type	Model	Start	End	Hours	Take Offs	Landings	
Signature									Totals to carry over

FLIGHT RECORD

YEAR		AIRCRAFT		FLIGHT TIME					GPS OR MAP LOCATION/ROUTE/REMARKS
									Totals Brought Forward
Month	Day	Type	Model	Start	End	Hours	Take Offs	Landings	
Signature									Totals to carry over

FLIGHT RECORD

YEAR		AIRCRAFT		FLIGHT TIME					GPS OR MAP LOCATION/ROUTE/REMARKS
									Totals Brought Forward
Month	Day	Type	Model	Start	End	Hours	Take Offs	Landings	
									Totals to carry over

Signature ____________________

FLIGHT RECORD

YEAR		AIRCRAFT		FLIGHT TIME					GPS OR MAP LOCATION/ROUTE/REMARKS
									Totals Brought Forward
Month	Day	Type	Model	Start	End	Hours	Take Offs	Landings	
Signature									Totals to carry over

YEAR		AIRCRAFT		FLIGHT TIME					GPS OR MAP LOCATION/ROUTE/REMARKS
									Totals Brought Forward
Month	Day	Type	Model	Start	End	Hours	Take Offs	Landings	
Signature									Totals to carry over

FLIGHT RECORD

YEAR		AIRCRAFT		FLIGHT TIME					GPS OR MAP LOCATION/ROUTE/REMARKS
									Totals Brought Forward
Month	Day	Type	Model	Start	End	Hours	Take Offs	Landings	
Signature									Totals to carry over

FLIGHT RECORD

YEAR		AIRCRAFT		FLIGHT TIME					GPS OR MAP LOCATION/ROUTE/REMARKS
									Totals Brought Forward
Month	Day	Type	Model	Start	End	Hours	Take Offs	Landings	
									Totals to carry over

Signature ______________________

FLIGHT RECORD

YEAR		AIRCRAFT		FLIGHT TIME					GPS OR MAP LOCATION/ROUTE/REMARKS
									Totals Brought Forward
Month	Day	Type	Model	Start	End	Hours	Take Offs	Landings	
Signature									Totals to carry over

FLIGHT RECORD

YEAR		AIRCRAFT		FLIGHT TIME					GPS OR MAP LOCATION/ROUTE/REMARKS
									Totals Brought Forward
Month	Day	Type	Model	Start	End	Hours	Take Offs	Landings	
									Totals to carry over

Signature ______________________

FLIGHT RECORD

YEAR		AIRCRAFT		FLIGHT TIME					GPS OR MAP LOCATION/ROUTE/REMARKS
									Totals Brought Forward
Month	Day	Type	Model	Start	End	Hours	Take Offs	Landings	
									Totals to carry over

Signature ____________________

FLIGHT RECORD

YEAR AIRCRAFT FLIGHT TIME GPS OR MAP LOCATION/ROUTE/REMARKS

									Totals Brought Forward
Month	Day	Type	Model	Start	End	Hours	Take Offs	Landings	
Signature									Totals to carry over

FLIGHT RECORD

YEAR		AIRCRAFT		FLIGHT TIME					GPS OR MAP LOCATION/ROUTE/REMARKS
									Totals Brought Forward
Month	Day	Type	Model	Start	End	Hours	Take Offs	Landings	
Signature									Totals to carry over

FLIGHT RECORD

YEAR		AIRCRAFT		FLIGHT TIME					GPS OR MAP LOCATION/ROUTE/REMARKS
									Totals Brought Forward
Month	Day	Type	Model	Start	End	Hours	Take Offs	Landings	
Signature									Totals to carry over

FLIGHT RECORD

YEAR		AIRCRAFT		FLIGHT TIME					GPS OR MAP LOCATION/ROUTE/REMARKS
									Totals Brought Forward
Month	Day	Type	Model	Start	End	Hours	Take Offs	Landings	
									Totals to carry over

Signature ______________________

FLIGHT RECORD

YEAR		AIRCRAFT		FLIGHT TIME					GPS OR MAP LOCATION/ROUTE/REMARKS
									Totals Brought Forward
Month	Day	Type	Model	Start	End	Hours	Take Offs	Landings	
									Totals to carry over

Signature ______________________

FLIGHT RECORD

YEAR		AIRCRAFT		FLIGHT TIME					GPS OR MAP LOCATION/ROUTE/REMARKS
									Totals Brought Forward
Month	Day	Type	Model	Start	End	Hours	Take Offs	Landings	
Signature									Totals to carry over

YEAR		AIRCRAFT		FLIGHT TIME					GPS OR MAP LOCATION/ROUTE/REMARKS
									Totals Brought Forward
Month	Day	Type	Model	Start	End	Hours	Take Offs	Landings	
Signature									Totals to carry over

FLIGHT RECORD

YEAR		AIRCRAFT		FLIGHT TIME					GPS OR MAP LOCATION/ROUTE/REMARKS
									Totals Brought Forward
Month	Day	Type	Model	Start	End	Hours	Take Offs	Landings	
Signature									Totals to carry over

FLIGHT RECORD

YEAR		AIRCRAFT		FLIGHT TIME					GPS OR MAP LOCATION/ROUTE/REMARKS
									Totals Brought Forward
Month	Day	Type	Model	Start	End	Hours	Take Offs	Landings	
									Totals to carry over

Signature ______________________

FLIGHT RECORD

YEAR		AIRCRAFT		FLIGHT TIME					GPS OR MAP LOCATION/ROUTE/REMARKS
									Totals Brought Forward
Month	Day	Type	Model	Start	End	Hours	Take Offs	Landings	
Signature									Totals to carry over

FLIGHT RECORD

YEAR		AIRCRAFT		FLIGHT TIME					GPS OR MAP LOCATION/ROUTE/REMARKS
									Totals Brought Forward
Month	Day	Type	Model	Start	End	Hours	Take Offs	Landings	
									Totals to carry over

Signature ______________________

FLIGHT RECORD

YEAR		AIRCRAFT		FLIGHT TIME					GPS OR MAP LOCATION/ROUTE/REMARKS
									Totals Brought Forward
Month	Day	Type	Model	Start	End	Hours	Take Offs	Landings	
									Totals to carry over

Signature ______________________

YEAR		AIRCRAFT		FLIGHT TIME					GPS OR MAP LOCATION/ROUTE/REMARKS
									Totals Brought Forward
Month	Day	Type	Model	Start	End	Hours	Take Offs	Landings	
Signature									Totals to carry over

FLIGHT RECORD

YEAR		AIRCRAFT		FLIGHT TIME					GPS OR MAP LOCATION/ROUTE/REMARKS
									Totals Brought Forward
Month	Day	Type	Model	Start	End	Hours	Take Offs	Landings	
Signature									Totals to carry over

FLIGHT RECORD

YEAR		AIRCRAFT		FLIGHT TIME					GPS OR MAP LOCATION/ROUTE/REMARKS
									Totals Brought Forward
Month	Day	Type	Model	Start	End	Hours	Take Offs	Landings	
									Totals to carry over

Signature ______________________

FLIGHT RECORD

YEAR		AIRCRAFT		FLIGHT TIME					GPS OR MAP LOCATION/ROUTE/REMARKS
									Totals Brought Forward
Month	Day	Type	Model	Start	End	Hours	Take Offs	Landings	
Signature									Totals to carry over

FLIGHT RECORD

YEAR		AIRCRAFT		FLIGHT TIME					GPS OR MAP LOCATION/ROUTE/REMARKS
									Totals Brought Forward
Month	Day	Type	Model	Start	End	Hours	Take Offs	Landings	
Signature									Totals to carry over

FLIGHT RECORD

YEAR		AIRCRAFT		FLIGHT TIME					GPS OR MAP LOCATION/ROUTE/REMARKS
									Totals Brought Forward
Month	Day	Type	Model	Start	End	Hours	Take Offs	Landings	
									Totals to carry over

Signature ______________________

YEAR AIRCRAFT FLIGHT TIME GPS OR MAP LOCATION/ROUTE/REMARKS

Month	Day	Type	Model	Start	End	Hours	Take Offs	Landings	Totals Brought Forward
Signature									Totals to carry over

FLIGHT RECORD

YEAR		AIRCRAFT		FLIGHT TIME					GPS OR MAP LOCATION/ROUTE/REMARKS
									Totals Brought Forward
Month	Day	Type	Model	Start	End	Hours	Take Offs	Landings	
Signature									Totals to carry over

FLIGHT RECORD

YEAR		AIRCRAFT		FLIGHT TIME					GPS OR MAP LOCATION/ROUTE/REMARKS
									Totals Brought Forward
Month	Day	Type	Model	Start	End	Hours	Take Offs	Landings	
									Totals to carry over

Signature ______________________

FLIGHT RECORD

YEAR		AIRCRAFT		FLIGHT TIME					GPS OR MAP LOCATION/ROUTE/REMARKS
									Totals Brought Forward
Month	Day	Type	Model	Start	End	Hours	Take Offs	Landings	
									Totals to carry over

Signature ______________________

FLIGHT RECORD

YEAR		AIRCRAFT		FLIGHT TIME					GPS OR MAP LOCATION/ROUTE/REMARKS
									Totals Brought Forward
Month	Day	Type	Model	Start	End	Hours	Take Offs	Landings	
									Totals to carry over

Signature ______________________

FLIGHT RECORD

YEAR		AIRCRAFT		FLIGHT TIME					GPS OR MAP LOCATION/ROUTE/REMARKS
									Totals Brought Forward
Month	Day	Type	Model	Start	End	Hours	Take Offs	Landings	
Signature									Totals to carry over

YEAR | AIRCRAFT | FLIGHT TIME | GPS OR MAP LOCATION/ROUTE/REMARKS

Month	Day	Type	Model	Start	End	Hours	Take Offs	Landings	Totals Brought Forward
Signature									Totals to carry over

FLIGHT RECORD

YEAR		AIRCRAFT		FLIGHT TIME					GPS OR MAP LOCATION/ROUTE/REMARKS
									Totals Brought Forward
Month	Day	Type	Model	Start	End	Hours	Take Offs	Landings	
Signature									Totals to carry over

FLIGHT RECORD

YEAR		AIRCRAFT		FLIGHT TIME					GPS OR MAP LOCATION/ROUTE/REMARKS
									Totals Brought Forward
Month	Day	Type	Model	Start	End	Hours	Take Offs	Landings	
Signature									Totals to carry over

FLIGHT RECORD

YEAR		AIRCRAFT		FLIGHT TIME					GPS OR MAP LOCATION/ROUTE/REMARKS
									Totals Brought Forward
Month	Day	Type	Model	Start	End	Hours	Take Offs	Landings	
Signature									Totals to carry over

FLIGHT RECORD

YEAR		AIRCRAFT		FLIGHT TIME					GPS OR MAP LOCATION/ROUTE/REMARKS
									Totals Brought Forward
Month	Day	Type	Model	Start	End	Hours	Take Offs	Landings	
Signature									Totals to carry over

FLIGHT RECORD

YEAR		AIRCRAFT		FLIGHT TIME					GPS OR MAP LOCATION/ROUTE/REMARKS
									Totals Brought Forward
Month	Day	Type	Model	Start	End	Hours	Take Offs	Landings	
Signature									Totals to carry over

YEAR		AIRCRAFT		FLIGHT TIME					GPS OR MAP LOCATION/ROUTE/REMARKS
									Totals Brought Forward
Month	Day	Type	Model	Start	End	Hours	Take Offs	Landings	
Signature									Totals to carry over

FLIGHT RECORD

YEAR		AIRCRAFT		FLIGHT TIME					GPS OR MAP LOCATION/ROUTE/REMARKS
									Totals Brought Forward
Month	Day	Type	Model	Start	End	Hours	Take Offs	Landings	
Signature									Totals to carry over

FLIGHT RECORD

YEAR		AIRCRAFT		FLIGHT TIME					GPS OR MAP LOCATION/ROUTE/REMARKS
									Totals Brought Forward
Month	Day	Type	Model	Start	End	Hours	Take Offs	Landings	
Signature									Totals to carry over

FLIGHT RECORD

YEAR		AIRCRAFT		FLIGHT TIME					GPS OR MAP LOCATION/ROUTE/REMARKS
									Totals Brought Forward
Month	Day	Type	Model	Start	End	Hours	Take Offs	Landings	
									Totals to carry over

Signature ______________________

FLIGHT RECORD

YEAR		AIRCRAFT		FLIGHT TIME					GPS OR MAP LOCATION/ROUTE/REMARKS
									Totals Brought Forward
Month	Day	Type	Model	Start	End	Hours	Take Offs	Landings	
									Totals to carry over

Signature ____________________

FLIGHT RECORD

YEAR		AIRCRAFT		FLIGHT TIME					GPS OR MAP LOCATION/ROUTE/REMARKS
									Totals Brought Forward
Month	Day	Type	Model	Start	End	Hours	Take Offs	Landings	
									Totals to carry over

Signature ____________________

YEAR		AIRCRAFT		FLIGHT TIME					GPS OR MAP LOCATION/ROUTE/REMARKS
									Totals Brought Forward
Month	Day	Type	Model	Start	End	Hours	Take Offs	Landings	
Signature									Totals to carry over

FLIGHT RECORD

YEAR | AIRCRAFT | FLIGHT TIME | GPS OR MAP LOCATION/ROUTE/REMARKS

									Totals Brought Forward
Month	Day	Type	Model	Start	End	Hours	Take Offs	Landings	
Signature									Totals to carry over

FLIGHT RECORD

YEAR		AIRCRAFT		FLIGHT TIME					GPS OR MAP LOCATION/ROUTE/REMARKS
									Totals Brought Forward
Month	Day	Type	Model	Start	End	Hours	Take Offs	Landings	
Signature									Totals to carry over

FLIGHT RECORD

YEAR		AIRCRAFT		FLIGHT TIME					GPS OR MAP LOCATION/ROUTE/REMARKS
									Totals Brought Forward
Month	Day	Type	Model	Start	End	Hours	Take Offs	Landings	
									Totals to carry over

Signature ______________________

FLIGHT RECORD

YEAR		AIRCRAFT		FLIGHT TIME					GPS OR MAP LOCATION/ROUTE/REMARKS
									Totals Brought Forward
Month	Day	Type	Model	Start	End	Hours	Take Offs	Landings	
									Totals to carry over

Signature ____________________

FLIGHT RECORD

YEAR		AIRCRAFT		FLIGHT TIME					GPS OR MAP LOCATION/ROUTE/REMARKS
									Totals Brought Forward
Month	Day	Type	Model	Start	End	Hours	Take Offs	Landings	
									Totals to carry over

Signature ____________________

YEAR		AIRCRAFT		FLIGHT TIME					GPS OR MAP LOCATION/ROUTE/REMARKS
									Totals Brought Forward
Month	Day	Type	Model	Start	End	Hours	Take Offs	Landings	
Signature									Totals to carry over

FLIGHT RECORD

YEAR		AIRCRAFT		FLIGHT TIME					GPS OR MAP LOCATION/ROUTE/REMARKS
									Totals Brought Forward
Month	Day	Type	Model	Start	End	Hours	Take Offs	Landings	
Signature									Totals to carry over

FLIGHT RECORD

YEAR		AIRCRAFT		FLIGHT TIME					GPS OR MAP LOCATION/ROUTE/REMARKS
									Totals Brought Forward
Month	Day	Type	Model	Start	End	Hours	Take Offs	Landings	
									Totals to carry over

Signature ____________________

FLIGHT RECORD

YEAR		AIRCRAFT		FLIGHT TIME					GPS OR MAP LOCATION/ROUTE/REMARKS
									Totals Brought Forward
Month	Day	Type	Model	Start	End	Hours	Take Offs	Landings	
Signature									Totals to carry over

FLIGHT RECORD

YEAR		AIRCRAFT		FLIGHT TIME					GPS OR MAP LOCATION/ROUTE/REMARKS
									Totals Brought Forward
Month	Day	Type	Model	Start	End	Hours	Take Offs	Landings	
Signature									Totals to carry over

FLIGHT RECORD

YEAR		AIRCRAFT		FLIGHT TIME					GPS OR MAP LOCATION/ROUTE/REMARKS
									Totals Brought Forward
Month	Day	Type	Model	Start	End	Hours	Take Offs	Landings	
									Totals to carry over

Signature ____________________

YEAR		AIRCRAFT		FLIGHT TIME					GPS OR MAP LOCATION/ROUTE/REMARKS
									Totals Brought Forward
Month	Day	Type	Model	Start	End	Hours	Take Offs	Landings	
Signature									Totals to carry over

FLIGHT RECORD

YEAR		AIRCRAFT		FLIGHT TIME					GPS OR MAP LOCATION/ROUTE/REMARKS
									Totals Brought Forward
Month	Day	Type	Model	Start	End	Hours	Take Offs	Landings	
Signature									Totals to carry over

FLIGHT RECORD

YEAR		AIRCRAFT		FLIGHT TIME					GPS OR MAP LOCATION/ROUTE/REMARKS
									Totals Brought Forward
Month	Day	Type	Model	Start	End	Hours	Take Offs	Landings	
Signature									Totals to carry over

FLIGHT RECORD

YEAR		AIRCRAFT		FLIGHT TIME					GPS OR MAP LOCATION/ROUTE/REMARKS
									Totals Brought Forward
Month	Day	Type	Model	Start	End	Hours	Take Offs	Landings	
									Totals to carry over

Signature ____________________

FLIGHT RECORD

YEAR		AIRCRAFT		FLIGHT TIME					GPS OR MAP LOCATION/ROUTE/REMARKS
									Totals Brought Forward
Month	Day	Type	Model	Start	End	Hours	Take Offs	Landings	
									Totals to carry over

Signature ______________________

FLIGHT RECORD

YEAR		AIRCRAFT		FLIGHT TIME					GPS OR MAP LOCATION/ROUTE/REMARKS
									Totals Brought Forward
Month	Day	Type	Model	Start	End	Hours	Take Offs	Landings	
									Totals to carry over

Signature ____________________

YEAR		AIRCRAFT		FLIGHT TIME					GPS OR MAP LOCATION/ROUTE/REMARKS
									Totals Brought Forward
Month	Day	Type	Model	Start	End	Hours	Take Offs	Landings	
Signature									Totals to carry over

FLIGHT RECORD

YEAR		AIRCRAFT		FLIGHT TIME					GPS OR MAP LOCATION/ROUTE/REMARKS
									Totals Brought Forward
Month	Day	Type	Model	Start	End	Hours	Take Offs	Landings	
Signature									Totals to carry over

FLIGHT RECORD

YEAR		AIRCRAFT		FLIGHT TIME					GPS OR MAP LOCATION/ROUTE/REMARKS
									Totals Brought Forward
Month	Day	Type	Model	Start	End	Hours	Take Offs	Landings	
Signature									Totals to carry over

FLIGHT RECORD

YEAR		AIRCRAFT		FLIGHT TIME					GPS OR MAP LOCATION/ROUTE/REMARKS
									Totals Brought Forward
Month	Day	Type	Model	Start	End	Hours	Take Offs	Landings	
									Totals to carry over

Signature ______________________

FLIGHT RECORD

YEAR		AIRCRAFT		FLIGHT TIME					GPS OR MAP LOCATION/ROUTE/REMARKS
									Totals Brought Forward
Month	Day	Type	Model	Start	End	Hours	Take Offs	Landings	
									Totals to carry over

Signature ....

FLIGHT RECORD

YEAR		AIRCRAFT		FLIGHT TIME					GPS OR MAP LOCATION/ROUTE/REMARKS
									Totals Brought Forward
Month	Day	Type	Model	Start	End	Hours	Take Offs	Landings	
									Totals to carry over

Signature ______________________

YEAR		AIRCRAFT		FLIGHT TIME					GPS OR MAP LOCATION/ROUTE/REMARKS
									Totals Brought Forward
Month	Day	Type	Model	Start	End	Hours	Take Offs	Landings	
									Totals to carry over

Signature ……………………………………

FLIGHT RECORD

YEAR		AIRCRAFT		FLIGHT TIME					GPS OR MAP LOCATION/ROUTE/REMARKS
									Totals Brought Forward
Month	Day	Type	Model	Start	End	Hours	Take Offs	Landings	
Signature									Totals to carry over

FLIGHT RECORD

YEAR		AIRCRAFT		FLIGHT TIME					GPS OR MAP LOCATION/ROUTE/REMARKS
									Totals Brought Forward
Month	Day	Type	Model	Start	End	Hours	Take Offs	Landings	
									Totals to carry over

Signature ______________________

FLIGHT RECORD

YEAR		AIRCRAFT		FLIGHT TIME					GPS OR MAP LOCATION/ROUTE/REMARKS
									Totals Brought Forward
Month	Day	Type	Model	Start	End	Hours	Take Offs	Landings	
									Totals to carry over

Signature ______________________

FLIGHT RECORD

YEAR		AIRCRAFT		FLIGHT TIME					GPS OR MAP LOCATION/ROUTE/REMARKS
									Totals Brought Forward
Month	Day	Type	Model	Start	End	Hours	Take Offs	Landings	
Signature									Totals to carry over

FLIGHT RECORD

YEAR		AIRCRAFT		FLIGHT TIME					GPS OR MAP LOCATION/ROUTE/REMARKS
									Totals Brought Forward
Month	Day	Type	Model	Start	End	Hours	Take Offs	Landings	
									Totals to carry over

Signature ______________________

YEAR		AIRCRAFT		FLIGHT TIME					GPS OR MAP LOCATION/ROUTE/REMARKS
									Totals Brought Forward
Month	Day	Type	Model	Start	End	Hours	Take Offs	Landings	
Signature									Totals to carry over

FLIGHT RECORD

YEAR		AIRCRAFT		FLIGHT TIME					GPS OR MAP LOCATION/ROUTE/REMARKS
									Totals Brought Forward
Month	Day	Type	Model	Start	End	Hours	Take Offs	Landings	
Signature									Totals to carry over

FLIGHT RECORD

YEAR		AIRCRAFT		FLIGHT TIME					GPS OR MAP LOCATION/ROUTE/REMARKS
									Totals Brought Forward
Month	Day	Type	Model	Start	End	Hours	Take Offs	Landings	
									Totals to carry over

Signature ....

FLIGHT RECORD

YEAR		AIRCRAFT		FLIGHT TIME					GPS OR MAP LOCATION/ROUTE/REMARKS
									Totals Brought Forward
Month	Day	Type	Model	Start	End	Hours	Take Offs	Landings	
									Totals to carry over

Signature ____________________

FLIGHT RECORD

YEAR		AIRCRAFT		FLIGHT TIME					GPS OR MAP LOCATION/ROUTE/REMARKS
									Totals Brought Forward
Month	Day	Type	Model	Start	End	Hours	Take Offs	Landings	
									Totals to carry over

Signature ______________________

FLIGHT RECORD

YEAR		AIRCRAFT		FLIGHT TIME					GPS OR MAP LOCATION/ROUTE/REMARKS
									Totals Brought Forward
Month	Day	Type	Model	Start	End	Hours	Take Offs	Landings	
									Totals to carry over

Signature ______________________

YEAR AIRCRAFT FLIGHT TIME GPS OR MAP LOCATION/ROUTE/REMARKS

Month	Day	Type	Model	Start	End	Hours	Take Offs	Landings	
									Totals Brought Forward
									Totals to carry over

Signature ______________________

FLIGHT RECORD

YEAR		AIRCRAFT		FLIGHT TIME					GPS OR MAP LOCATION/ROUTE/REMARKS
									Totals Brought Forward
Month	Day	Type	Model	Start	End	Hours	Take Offs	Landings	
Signature									Totals to carry over

FLIGHT RECORD

YEAR		AIRCRAFT		FLIGHT TIME					GPS OR MAP LOCATION/ROUTE/REMARKS
									Totals Brought Forward
Month	Day	Type	Model	Start	End	Hours	Take Offs	Landings	
Signature									Totals to carry over

FLIGHT RECORD

YEAR		AIRCRAFT		FLIGHT TIME					GPS OR MAP LOCATION/ROUTE/REMARKS
									Totals Brought Forward
Month	Day	Type	Model	Start	End	Hours	Take Offs	Landings	
									Totals to carry over

Signature ______________________

FLIGHT RECORD

YEAR		AIRCRAFT		FLIGHT TIME					GPS OR MAP LOCATION/ROUTE/REMARKS
									Totals Brought Forward
Month	Day	Type	Model	Start	End	Hours	Take Offs	Landings	
									Totals to carry over

Signature ______________________

FLIGHT RECORD

YEAR		AIRCRAFT		FLIGHT TIME					GPS OR MAP LOCATION/ROUTE/REMARKS
									Totals Brought Forward
Month	Day	Type	Model	Start	End	Hours	Take Offs	Landings	
									Totals to carry over

Signature ____________________

YEAR		AIRCRAFT		FLIGHT TIME					GPS OR MAP LOCATION/ROUTE/REMARKS
									Totals Brought Forward
Month	Day	Type	Model	Start	End	Hours	Take Offs	Landings	
Signature									Totals to carry over

Made in the USA
Monee, IL
04 November 2026

67380092R10059